23rd Topical Conference on Gas Utilization 2023

Topical Conference at the 2023 AIChE Spring Meeting and 19th Global Congress on Process Safety

Houston, Texas, USA
12-16 March 2023

ISBN: 978-1-7138-7930-5

Printed from e-media with permission by:

Curran Associates, Inc.
57 Morehouse Lane
Red Hook, NY 12571

Some format issues inherent in the e-media version may also appear in this print version.

Printed with permission by Curran Associates, Inc. (2023)

For permission requests, please contact AIChE
at the address below.

AIChE
120 Wall Street, FL 23
New York, NY 10005-4020

Phone: (800) 242-4363
Fax: (203) 775-5177

www.aiche.org

Additional copies of this publication are available from:

Curran Associates, Inc.
57 Morehouse Lane
Red Hook, NY 12571 USA
Phone: 845-758-0400
Fax: 845-758-2633
Email: curran@proceedings.com
Web: www.proceedings.com

TABLE OF CONTENTS

21a Decarbonizing the Natural Gas Processing Plants for Hydrogen, and Syngas Production: The Potentials, and the Challenges

Mohamed Sufiyan Challiwala, Hanif Choudhury, Gasim Ibrahim, Nimir Elbashir

At 8.5 GtCO2 in 2020, industrial greenhouse gas (GHG) emissions account for approximately 24% of total global emissions. Therefore, the industrial sector will have to reduce emissions by 1.2% to 7.4 GtCO2 by 2030 to achieve sustainability commitments. Thus, the decarbonization of industries using novel solutions coupled with accelerated use of renewable energy at higher efficiency has been a trend-setting research topic. To this end, natural gas utilization has received tremendous global interest due to being a cleaner alternative and its ability to offer flexibility for use in heating and power generation, as well as acting as a precursor for many chemical products. Furthermore, natural gas-based industrial technologies typically have lower capital costs, operating costs, and electricity consumption than benchmark processes. Natural gas use has also seen significant impetus due to its tremendous availability from cheaper sources, such as shale gas in the United States and natural gas from the middle east. It should be noted that at least 0.2 kg CO2e GHGs are emitted per kg of natural gas produced from Shale gas and conventional recovery processes (as per the GREET 2021 database). This is tremendous, especially when natural gas production is increased for the energy transition. Several proven approaches for handling CO2 emissions from the upstream and the midstream process plants have been developed, such as enhanced oil recovery and geological storage techniques. Nevertheless, stricter environmental regulations and uncertainties pertaining to the re-emission of CO2 from geological sites have encouraged the adoption of chemical conversion techniques.

This presentation will discuss the potential natural gas in the hydrogen era and the future of LNG as part of the global energy mix. Decarbonization of the aforementioned sectors is necessary for their continuing contribution to the energy demand. Our discussion will cover several examples of decarbonizing the industry, including our CARGEN® Technology. The novel CARGEN® process converts industrial GHG emissions to solid carbon material and hydrogen-rich syngas (which is a valuable chemical intermediate for gas to liquid [GTL] processing). More importantly, CARGEN® Technology presents a desirable benefit of decarbonizing Steam Methane Reforming (SMR) process that has a footprint of ~10 kg CO2e/kgH2 by producing almost zero net carbon footprint hydrogen. It should be noted the SMR, and methane decomposition processes together account for more than 90% of global industrial hydrogen production. Thereby, CARGEN® Technology not only presents a centralized decarbonization link between hydrogen production and carbon nanotube production but also offers a substantial competitive advantage of low CAPEX introduction in the existing industries by debottlenecking approach.

References:

1. Challiwala, Mohamed S., et al. 'A novel CO_2 utilization technology for the synergistic co-production of multi-walled carbon nanotubes and syngas.' Scientific reports 11.1 (2021): 1-8.
2. Challiwala, Mohamed S., et al. 'Turning CO_2 into Carbon Nanotubes' Chemical Engineering Progress, October (2020) issue.
3. Challiwala, Mohamed, et al. 'Enhanced CO_2 fixation and Syngas production using a two-reactor setup'. US20200109050A1; AU2018249486B2; PCT/US2018/025696; WO2018187213
4. Challiwala, Mohamed S., et al. 'Regeneration and Activation of Catalyst for Carbon and Syngas Production'. PCT/QA2020/050012; CN114340789; WO/2020/185107; IN202117046468
5. Challiwala, Mohamed S., et al. 'Novel Catalyst for CARGEN™ Process and a method to prepare the same." PCT/QA2020/050012; EP20202903512; WO2021125990

57a Toward Chemical and Thermal Synergies of Integrating Methane Oxidative Coupling and Dehydroaromatization across the Scales

Mamoun Al-Rawashdeh, Maria Haki, Raid J. Hassiba, Izabel Medeiros-Costa, Patrick Linke

Methane dehydroaromatization (MDA) is a one-step reaction that converts methane into aromatics and hydrogen with zero CO_2 emission from the reaction stoichiometry. Oxidative coupling of methane (OCM) is a one-step reaction that converts methane into valuable C_2+ products along with CO_2 as the main by-product. Although very attractive reactions, they are not yet commercialized due to multiple technical challenges. MDA is faced with the challenge of fast catalyst deactivation, thermodynamic limitations that limit maximum achieved conversion, and source of heating as the reaction is endothermic and requires high temperature. OCM is highly exothermic which requires strict and difficult heat management and has a tradeoff between C2+ selectivity and methane conversion. This work is an effort to address these technical challenging by integrating these chemistries toward achieving chemical and thermal synergies. Different options of integration are considered starting at the catalyst scale, to the reactor, and then the process scale. A structured approach to address such a complex problem across the different scales will be presented. At the reactor scale, experimental investigations will be presented to provide evidence for OCM-MDA catalyst integration benefits and limitations in the same reactor. The aim is to demonstrate the opportunities and limitations for developing this promising process concept.

57b Review on the Integration Processes on Ethylene Oxide Derivatives; How the Dynamic of Operations Can Maximize the Company Revenue

Ammar Bakheet

Abstract:
High-value-added manufacturing companies deal with a competitive environment and market where The demands of material are varying specially in petrochemical sectors due to uncertainty in energy market, other reasons related to oil & gas and global inflation.

Traditional & limited manufacturing processes will limit the economic benefit. Hence, looking for production operational strategies for high-value-added manufacturing companies by maximizing the derivatives based on the global demands. The review covers the most possible Ethylene Oxide (EO) derivatives that can be obtained

In addition, the paper gives a comprehensive overview of the research advances achieved in the field of production operational strategies for high-value-added manufacturing companies in Ethylene Oxide (EO) derivatives , identifying key product as well as key of use and outlook which derived from a structured analysis, research ,safety standard, and demand

57c Carbon Dioxide Utilization in the Natural Gas Industry Via Direct Conversion to Organic Carbonates

Gasim Ibrahim, Hanif Choudhury, Mohamed Sufiyan Challiwala, Nimir Elbashir

Natural gas is considered by many to be a transitional fuel with significantly lower environmental impact compared to other energy sources such as coal and oil. However, the production and utilization of natural gas still come with significant greenhouse gas emissions, including CO_2, which has made the decarbonization of the natural gas industry through CO_2 capture, sequestration, and utilization (CCSU) the highest priority for the industry. While the idea of chemical utilization of CO_2 to produce other molecules is appealing, CO_2 is a stable molecule that is challenging to convert into added-value chemicals efficiently. One means of facilitating CO_2 reactions is via catalysis, wherein the CO_2 molecules are adsorbed on the catalyst's surface and interact with other precursors to form products. A challenging aspect of CO_2 reactions is low conversion due to thermodynamic constraints even in the presence of catalysts – as is observed in the conversion of CO_2 to Organic Carbonates.

This paper will highlight our efforts in developing a sustainable process for the direct synthesis of dimethyl carbonate (DMC), which is a flexible molecule used in a variety of applications such as a sustainable and green solvent, an electrolyte in the battery industry, a fuel additive, and in the manufacturing of engineering plastics. The direct synthesis of DMC involves reacting methanol and CO_2 to produce DMC and water as a means of decarbonizing the natural gas industry. We utilize reaction engineering concepts and techniques to synthesize, characterize, and develop catalysts and then test the efficacy of those catalysts for the direct synthesis of DMC. We synthesized catalysts using xero-gel method and co-precipitation techniques, and the performance of these catalysts was evaluated in a slurry reactor under the batch condition of 5-30 bar pressure and 110-130 °C temperature. The catalysts exhibited significantly different conversions towards DMC, but all below 1% conversion – in line with our thermodynamic assessment (Elbashir, Choudhury and Challiwala, 2021). Furthermore, a metric we developed to study the potential for CO_2 fixation through chemical reactions indicated the process is CO_2 neutral and negative only when the reaction has a conversion above 30% when the reaction is carried out in these relatively low-pressure conditions. Thus, to approach this target, we attempt to shift the equilibrium conversion by utilizing methods for in-situ removal of water, which shifts the equilibrium in favor of the forward reaction. Our test with physical adsorbents such as activated carbon, Dowex, and 5A molecular sieves showed no improvement in the methanol conversion toward DMC. However, the addition of 2-CP as a water scavenger, even at a low-pressure reaction condition of 5 bar, showed significant improvement in methanol conversion towards DMC (Tomishige, Tamura and Nakagawa, 2018). An optimization of 2-CP system for low-pressure conditions of 5 bar is underway to approach the conversion needed to make the direct synthesis of DMC viable reaction pathways for CO_2 sequestration. Currently, we are conducting modeling, simulations, and kinetic analyses in preparation for testing the reaction in a relatively large scale (10 l/hr) Reactive Water Distillation unit.

This work is funded by a research grant from Qatar National Research Fund under its flagship NPRP program (Project No. NPRP13S-0122-200141) and co-funded by Shell Global Solutions (Qatar Shell Research Technology Center).

References

Elbashir, N. O., Choudhury, H. A. and Challiwala, M. (2021) 'Methods of Producing Dimethyl Carbonate'. World Intellectual Property Organization International Bureau.

Tomishige, K., Tamura, M. and Nakagawa, Y. (2018) 'CO_2 Conversion with Alcohols and Amines into Carbonates, Ureas, and Carbamates over CeO_2 Catalyst in the Presence and Absence of 2-Cyanopyridine', *The Chemical Record*. John Wiley & Sons, Ltd. doi: 10.1002/tcr.201800117.

74a CO2Fix: A Simple and Efficient Tool for Estimating the Decarbonization Potential for Chemical CO_2 Utilization Pathways

Gasim Ibrahim, Mohamed Sufiyan Challiwala, Hanif Choudhury, Mahmoud El-Halwagi, Nimir Elbashir

A global effort is underway to reduce GHG emissions to address the growing concerns of anthropogenic climate change. Carbon dioxide capture, utilization, and storage (CCUS) has garnered significant attention as a means of reducing GHG emissions and decarbonizing several industries, including the natural gas industry. One means of applying CCU to a process is through the chemical conversion of CO_2 produced from the process into other products. However, a given CCU process will have its CO_2 footprint, which has to be accounted for when analyzing the process's potential for CO_2 abatement or when comparing such potential to other processes. This is typically done by conducting Life Cycle Assessments (LCA), which involve comprehensive accounting for the CO_2 footprint of a process throughout its various stages.

In this work, we develop a new metric to account for the CO_2 and carbon footprint of chemical processes in addition to the CO_2-fixation probability for CO_2 utilization technologies. This new tool is named CO2Fix, defined as the ratio of CO_2 being fixed chemically from a process while accounting for the CO_2 produced directly or indirectly from the same process through its chemistry, energy requirements and/or other emissions. Leveraging basic thermodynamic concepts and ASPEN® aided regression analysis, a model was developed for estimating CO2Fix that is a function of easily obtainable reaction parameters such as temperature, pressure, the heat of reaction, CO_2 conversion, and CO_2 intensity of utilized energy sources. The developed model was then used to analyze the CO2Fix and understand its relationship with multiple model parameters over a wide range of reaction conditions using different energy sources (conventional and renewable). The analysis provided insights such as the nonlinear relationship between CO2Fix and the conversion for non-autothermal reactions since the relationship is normally linear for autothermal reactions. The model was then used in several case studies to analyze the CO2Fix of a few known CO_2 utilization processes, such as CO_2 hydrogenation to methanol (which is one of the few commercial CO_2 chemical utilization routes), the Dry Reforming of Methane (DRM) reaction, and the direct synthesis of Dimethyl Carbonate (DMC) using CO_2. In the case of CO_2 hydrogenation to methanol, the CO2Fix was calculated to be 1.6 at a CO_2 conversion of 50% when utilizing gas as an energy source. Analysis of the DRM reaction showed that the reaction has no potential to abate CO_2 (CO2Fix of 1 or below) without utilizing low CO_2 intensity renewable energy sources such as solar power or wind. Even when using biomass-powered electricity, the CO2Fix for the DRM reaction was approximately 1.1 at 100% conversion. The direct synthesis of DMC was found to have the potential for CO_2 abatement over a large range of reaction conditions at CO_2 conversions beyond 30%, giving a CO2Fix >1 even when the process is powered by natural gas. With the current emphasis on CCUS as a means of decarbonizing the natural gas industry, we propose this model as an efficient and straightforward approach to quickly screen out unviable CCUS routes. Furthermore, the CO2Fix model can be applied to compare different CCUS routes in terms of their ability to decarbonize processes through the chemical fixation of CO_2. Additionally, a software tool was developed utilizing the CO2Fix model for rapid analysis of CO_2 reactions with limited information.

74c SNG As Renewable Energy Transport Media

Yoshitsugi Kikkawa

In recent years, reducing CO2 emission has been the focus for addressing the concern over global warming and presumed related abnormal weather phenomena such as wildfire in California, US and Turkish, serious flooding in US, Europe and China, strong hurricanes, drought, and the melting of the permafrost of Alaska and Siberia and polar ice caps. The natural gas industry has already taken steps to reduce the CO2 emission by using high efficiency gas turbines instead of steam turbine systems. Will these steps be enough to satisfy the ever-increasing social concerns, or will further reductions be required?

To respond the social concern, following ideas are presented:

- Methane, SNG, synthesis: renewable energy, i.e., wind power generation lead to hydrogen production by electrolysis. Methane is synthesized by Sabatier reaction with recycled CO2.
- The SNG is compressed by compressor and transferred by pipeline
- Receiving Terminal and Power Generation: application of high efficiency combined cycle systems, CO2 recovery by amine process from gas turbine exhaust and CO2 is compressed for transport.
- The CO2 is compressed by compressor and transferred by pipeline and recycled to the SNG synthesis.

This concept would lead to higher power costs for the consumer as around 30 cents per kWh, it may look costly, however it will solve the global warming issue completely, which has been started since industrial revolution at 18th century in UK.

37a VCE Potential Cascade Effects – Part 2

Cynthia Spitzenberger, Mehmet Ozbey, Michael Johnson, Tatiana Norman, Gabriele Ferrara

The LNG industry has historically maintained a high degree of safety in its operations. This has been built on a clear understanding of the hazards associated with LNG processes, storage and transportation. A vapor cloud explosion (VCE) occurs following the ignition of significant vapor clouds formed by accidental release of flammable fluids and can involve generating damaging overpressures. LNG, which in its vapor form is natural gas, is inherently safer in this respect compared to other fuels, but this does not remove the need to consider these events.

A key aspect of a VCE is that the resulting damaging overpressures can cause further failures in equipment and as a result, the initial incident escalates. These 'Cascade Effects' can significantly increase the severity and extent of a VCE event. Understanding the potential for Cascade Effects on LNG facilities can inherently improve safety, protecting both people and key energy infrastructure. The US DOT PHMSA commissioned a research project to evaluate the severity of potential explosions and the response of structures to exceptional loads. DNV and Thornton Tomasetti partnered to conduct the research project starting in Q3 2020.

DNV performed dispersion and explosion modeling with CFD to evaluate the potential extent of explosion loads from flammable materials present at LNG facilities. Finite Element Analysis (FEA) modeling was performed by Thornton Tomasetti to evaluate the structural response of representative equipment to overpressure loads.

At the 2022 conference, intermediate results and the research to date were presented. The conference paper in 2023 will summarize the outcomes of the research and the resulting overall findings and conclusions based on the research. The research is meant to provide guidance on assessing the potential for Cascade Effects and potential mitigation options.

This work was funded in part, under the Department of Transportation, Pipeline and Hazardous Materials Safety Administration. The views and conclusions contained in this document are those of the authors and should not be interpreted as representing the official policies, either expressed or implied, of the Pipeline and Hazardous Materials Safety Administration, the Department of Transportation, or the U.S. Government.

Evaluation Protocol for Source Term Models

Filippo Gavelli
Blue Engineering and Consulting
6011 University Blvd, Suite 220, Ellicott City, MD 21043
fgavelli@blueeandc.com

Bryant Hendrickson
Blue Engineering and Consulting
6011 University Blvd, Suite 220, Ellicott City, MD 21043
bhendrickson@blueeandc.com

Keywords: model evaluation; model validation; accidental release; LNG

Abstract

The siting of LNG facilities requires several types of hazards to be quantified so the risk to the public and public property in areas surrounding each facility can be properly assessed. In order to quantify the hazards from potential accident scenarios, computational tools must be used. Given the multiplicity of tools available, ranging from empirical models to 3D computational fluid dynamics (CFD) packages, the agency reviewing the project application may not have the methodologies or protocols necessary to determine the suitability of a given model to a certain scenario, its accuracy, or any other setup requirements.

A critical step in assessing the consequences of a hazardous fluid release is the characterization of the "source term", that is, the quantification of the release rate and the conditions of the release. In fact, an error in the source term will propagate to the calculation of the consequences of the release (e.g., flammable dispersion, jet fires, etc.). Therefore, model evaluation needs to include the characterization of the source term.

This paper describes the Model Evaluation Protocol (MEP) developed for source term models used for LNG facility siting studies. The paper will discuss the evaluation methodology, including the information to be submitted by the model proponents and the review process, the data set to be used for model validation, and the criteria to evaluate each model and to define their respective validation factors (i.e., safety factors).

Introduction

The safe siting of LNG facilities requires project developers to evaluate the potential consequences to the public and public property from accidental loss of containment and release of hazardous materials (e.g., flammable and/or toxic). These consequences include: pool, jet or flash fires; vapor cloud explosions; toxic gas exposure; and catastrophic vessel failures. Calculating the distance to which hazardous conditions may extend for each type of scenario requires the use of computational models.

Assessing the quality of models, especially those used to evaluate the consequences of hazardous scenarios on the public, has long been a concern to regulatory bodies. Therefore, the Pipeline and Hazardous Materials Safety Administration (PHMSA) commissioned Blue Engineering and Consulting Company (BLUE) and the Gas Technology Institute (GTI), via research project #854 titled: "Develop an Evaluation Protocol for Non-LNG Release Hazards – Modeling", to develop a set of MEPs, that will allow the review of modeling tools for several different hazards, including flammable and toxic dispersion, source terms for dispersion models, vapor cloud explosions, and BLEVEs.

A critical step in assessing the consequences of a hazardous fluid release is the characterization of the "source term", that is, the quantification of the release rate and the conditions of the release. In fact, an error in the source term will propagate to the calculation of the consequences of the release (e.g., flammable dispersion, jet fires, etc.). Therefore, model evaluation needs to include the characterization of the source term. This paper provides a summary of the model evaluation protocol for source term models. The complete documents (respectively, the model evaluation protocol and the associated model validation database can be obtained from PHMSA's project website at: https://primis.phmsa.dot.gov/matrix/PrjHome.rdm?prj=854.

Scope of the Model Evaluation

The scope of the Source Term MEP is to define an objective and transparent methodology to determine the ability of a model to accurately quantify releases for scenarios typically evaluated in LNG facility siting. The protocol itself is intended to be a guidance document which outlines the tasks required by the petitioner and the reviewer. The purpose of the source term MEPs is to evaluate models that calculate the mass flow rate and other relevant characteristics (e.g., velocity, temperature, liquid fraction, etc.) of accidental releases. The objective of evaluation is to allow the regulatory body (i.e., PHMSA) to determine whether a given model should be approved or rejected for use in the calculation of the source term for accidental releases, within the context of LNG facility siting studies. Additionally, the evaluation should allow the regulatory body to specify any restrictions on the use of an approved model, as well as any "validation factors" to be applied to the modeling results. Comparing models against one another and establishing rankings of preferred models are not objectives of the MEP.

Types of Release Scenarios

When a loss of containment occurs in a pipe or vessel containing a hazardous fluid, different types of release are possible:

1) Liquid spills are liquid releases from low pressure storage (e.g., atmospheric LNG storage tanks) or from large-bore holes, which results in negligible flashing of the released stream. A liquid spill typically falls to the ground (although spills on water are also considered in some cases) and can either spread on the terrain, be bounded by a curb, or be conveyed to a dedicated collection area. Another scenario potentially resulting in a liquid spill is a pressurized liquid jet: in this case, liquid droplets that fall out of the jet stream or impinge on a solid surface will deposit on the ground and collect to form a liquid pool (this is discussed further in the next item).
2) Jet releases occur when a pipe or vessel, containing a fluid under pressure, is breached. If the breached system contains a pressurized liquid, stored above its normal boiling point, a two-phase release will occur as a portion of the liquid flashes to vapor while expanding to the atmosphere. Expansion of the jet to atmosphere occurs within a short distance from the orifice (on the order of a few orifice diameters); the jet then continues at atmospheric pressure. In the expansion region, the jet is also atomized to droplets; the expansion region is too small to interact appreciably with the atmosphere, therefore, no air is entrained into the jet in this region. Once the jet has expanded to atmosphere and continues to travel away from the source (the jet entrainment region), the jet entrains a mass of air many times that of the vapor in the jet, and the jet slows to a fraction of its initial velocity. Air entrainment is therefore an important phenomenon, which has a strong effect on the characteristics of the jet (temperature, liquid fraction, velocity, etc.). If the jet continues unimpeded, the liquid droplets progressively evaporate as more air is entrained; the heavier

droplets, however, may fall out of the jet stream and deposit onto the ground (i.e., rain-out), forming a liquid pool. If the jet encounters obstacles (e.g., a wall) while droplets are still present, part or all of the remaining liquid will deposit onto the solid surface and fall to the ground, also contributing to rain-out and the formation of a liquid pool. Eventually, through a combination of evaporation and rain-out, the liquid fraction of the jet reduces to zero leaving a single-phase jet: for some models (e.g., FLACS) this condition represents the jet pseudo-source (i.e., how the leak is defined in FLACS).

3) The term catastrophic release describes the instantaneous release of the inventory within a vessel, following the sudden rupture of the vessel itself. The rupture may occur "spontaneously" or as a result of exposure to elevated heat fluxes (e.g., from a pool fire). Vessel rupture due to fire exposure would result in a boiling liquid expanding vapor explosion (BLEVE), which will be the subject of a separate model evaluation. The spontaneous rupture of a storage vessel (a.k.a., Pressure Vessel Burst, PVB) is not considered credible under current US regulations for facility siting; however, vessel rupture has a nonzero probability of occurrence within the context of a risk assessment. The key parameters affecting the source term for a catastrophic release are the composition and physical state (i.e., pressure and temperature) of the material being stored. Very limited experimental data is available that evaluates catastrophic releases and, for the most part, it involves the instantaneous release of dense gases stored at near atmospheric conditions (e.g., Thorney Island). Therefore, catastrophic releases were not included in this database.

Source Term Models

Source term models for liquid spills currently fall into one of two categories:

1) Semi-empirical models: these models calculate the spreading and vaporization of a liquid spill assuming the liquid pool to be unconfined and spreading onto flat, unobstructed terrain. As a result, the liquid pool maintains a circular shape and the pool radius can be calculated as a function of time based on simplified mass and force balances. Several circular (or semicircular) pool spreading models have been developed which can be implemented by interested users in a spreadsheet or other computing tool or programming script. Additionally, several commonly used consequence models (e.g., Phast, CANARY, ALOHA, etc.) already include circular pool spreading models to provide the vapor generation term for dispersion or pool fire calculations.
2) Shallow-water models: these models calculate the spreading and vaporization of a liquid spill by solving the 2D shallow-water equations (SWEs). The SWEs are based on the assumption that the thickness of the liquid pool is much smaller than its other dimensions, therefore, the fluid properties can be approximated as uniform in the vertical (z) direction, leading to a two-dimensional system of equations. SWE models allow the spreading and vaporization to take into account terrain variations, obstacles, etc. User implementation of the SWEs is rather complicated, however, commonly used CFD consequence models (e.g., FLACS, KFX) already include SWE models to provide the vapor generation term for dispersion or pool fire calculations. Additionally, several tested implementations of the SWEs are publicly available (e.g., TwoDee, AnuGa), although they are typically aimed at water studies (e.g., flooding, tsunamis) and therefore do not include a vapor generation term.

For pressurized jet releases, the simplest model to calculate the release rate through a breach is the orifice flow model [6], which is based on Bernoulli's equation for steady, incompressible and inviscid flow, and is generally formulated as:

$$\dot{m} = C_d \, A \, f(fluid, P_i - P_{atm}) \quad (1)$$

where:

- C_d is the discharge coefficient, which depends on the fluid phase as well as the shape and conditions of the breach;
- A is the flow area of the breach;
- P_i is the pressure inside the system;
- P_{atm} is the atmospheric pressure;
- f is a function whose formulation varies depending on the pressure difference and fluid properties inside the system.

The orifice flow model is often included in consequence modeling tools (e.g., dispersion or fire) or can be easily implemented by users through a spreadsheet or other calculator to quantify the source term for hazard modeling. A limitation of the orifice flow model is the assumption of constant fluid properties, which may become significant when the release fluid is near saturation or in superheated liquid conditions: in these cases, phase change will occur during the release (in some cases it may even occur inside the pressurized system) and that may affect the flow rate through the breach as well as the conditions of the emerging stream. It is important to note that models utilizing Bernoulli's equation and the liquid properties at storage conditions are considered conservative and do not require to be evaluated against experimental data sets for liquid jet releases. The same applies to the formulas for the release of pressurized gas jets, both subsonic and choked.

More sophisticated models for the flow through an orifice have been developed and implemented in consequence modeling tools (e.g., Phast) to take into account the variability of thermodynamic properties of the fluid as it flows through the breach and expands to atmosphere. The most common expansion models are based on the assumption of the expansion to atmosphere being either isenthalpic or isentropic.

Computational fluid dynamics (CFD) models may also be used to simulate the fluid flow through a breach. CFD models can provide a much higher level of detail than models based on mass, momentum, and energy balance; however, the improvement in the source term definition, currently, does not justify the level of computational effort required to set up and run CFD models for flow releases.

The Model Evaluation Protocol

The scope of the source term MEP is to define an objective and transparent methodology to determine the ability of a model to accurately quantify dispersion hazards in the context of LNG facility siting. The MEPs are guidance documents which outline the tasks required by the model petitioner and the reviewer. Specifically, they provide guidance to the petitioner regarding:

- Information to be submitted (e.g., user guide, technical references, etc.);
- Scenarios to be simulated for validation (including any sensitivity analysis); and
- Feedback that should be expected once the evaluation is completed (e.g., whether the model is accepted or not, restrictions on the applicability of the model, or validation factors to be applied to the modeling results).

The protocol also provides guidance to the evaluator, including:

- Requirements that an evaluator must meet to be qualified;
- Tasks to be performed during the evaluation; and
- Reporting requirements.

The source term MEP follows the general model evaluation methodology described by Gavelli et al. [3] and includes the following tasks:

1) Scope of the evaluation
2) Evaluator qualifications
3) Model description
4) Scientific assessment
5) User-oriented assessment
6) Verification
7) Validation, which includes:
    a) Model Validation Database, with:
        i) Datasets for model validation
        ii) Key variables and physical comparison parameters
        iii) Model performance measures
    b) Sensitivity analysis
    c) Qualitative performance assessment
    d) Quantitative performance assessment
8) Model Evaluation Report, which includes:
    a) Summary of the evaluation (i.e., steps 4 through 7 above)
    b) Best practice guidance
    c) Recommendations for improvement

For each MEP, the model validation (step 7 above) is included in a separate document – the Model Validation Database. This allows the validation data sets, model performance measures and acceptability criteria to be updated as more information becomes available (e.g., new experimental data sets or results from model evaluations) without requiring a complete revision of the model evaluation protocol.

The Model Validation Database

Model validation is the process of comparing the predictions of a model against experimental data relevant to the types of scenarios that the model is intended to predict. In the context of a model evaluation protocol to be used for regulatory purposes, the term "validation" is not intended in absolute terms but, rather, it implies "relative to the scenarios against which it was compared". Therefore, the purpose of a model validation database is to identify a set of scenarios to be simulated with a computational model, in such a manner that qualitative and/or quantitative comparisons may be made between model predictions and actual observations. In addition to specifying data sets for model validation, the MVD needs to include the information necessary to set up and perform the simulations, as well as define the criteria for comparison of the model predictions with the observations.

The following criteria were applied to datasets reviewed for inclusion in the source term MVD:

- The test configuration (i.e., material, pressure, temperature) should be representative of realistic scenarios. The scale of the experiment (i.e., release rate) should also be comparable to actual scenarios.
- The test series should include a wide range of release conditions.
- The tests should be described in sufficient detail for modelers to set up their simulations in an accurate and consistent manner, with as few assumptions as possible.
- Ambient conditions (i.e., temperature, humidity, wind speed and atmospheric stability, etc.) should be available from sensors in/near the area of interest. For trials involving the evaporation of liquid spills, instrument resolution should be sufficient to define the atmospheric boundary layer profiles.

- Sensor data should include measurement of the key variables. In the case of source term experiments, release flow rate measurements must be available. Liquid rainout measurements are also critical in the case of flashing liquid releases.
- Measurements should be of a sufficient quantity to be statistically representative.
- Uncertainty of all measured and derived quantities should be provided, together with a description of the method used to define such uncertainty.
- If the raw data is not provided, the time averaging applied to the data must be specified.
- The data must be freely available.

In order to optimize the database, the test series were selected to span a broad range of conditions and to strike a balance between the number of trials included for each type of scenario. For that purpose, the test series were grouped according to several parameters as follows:

- Release Type:
  - Jet
  - Spill
- Release Phase:
  - Liquid
  - Gas/Vapor
- Spill substrate (for spill releases only):
  - Solid
  - Water

Numerous test series were reviewed for the source term MVD; each series was "rated" on a 5-point scale according to several categories, following an approach first introduced by Skjold et al. [1] and previously applied to the flammable and toxic dispersion MEPs ([2], [3], [4]). Table 1 lists the test series that were included in the database and Table 2 shows them grouped by type of release (rows) and release phase (columns). A summary of each test series included in the MVD can be found in the source term MVD report on PHMSA's project website.

Table 1. Experimental test series included in the Source Term MVD.

Series	Fluid	Release Type	Release Phase [Substrate, for Spills]
Bureau of Mines [9]	LNG	Spill	Water
Burro [10]	LNG	Spill	Water
CO2PIPETRANS [11], [12]	Carbon Dioxide	Jet	Gas
CCPS Rainout Exp. [13]	Water; Chlorine	Jet	Liquid
Desert Tortoise [14]	Ammonia	Jet	Liquid
Goldfish [15], [16]	Hydrofluoric Acid	Jet	Liquid
Goldwire et al. [17]	Ammonia	Spill	Ground
Jack Rabbit II [18]	Chlorine	Jet	Liquid
Lathen (EEC) [19]	Propane	Jet	Liquid
Reid et al. [20]	LPG	Spill	Water
Univ. of Sci. and Tech. (Korea) [21], [22]	Nitrogen	Spill	Concrete

Table 2. Source Term MVD test series, grouped by Release Type and Phase.

	Jet	Spill
Liquid	CCPS Rainout Desert Tortoise Goldfish Jack Rabbit II Lathen (EEC)	Bureau of Mines Burro Goldiwre et al. Reid et al. Univ. of Science and Tech. (Korea)
Gas	CO2PIPETRANS	N/A

The need for multiple experimental data sets is due to the complex physics involved in the release of liquid streams, particularly when stored under pressure at temperatures above their normal boiling points, and the numerous models that can be used to calculate the release parameters. Additionally, in some cases, test series (e.g., Desert Tortoise, Goldfish) were included in the source term MVD because they had been previously included in at least one of the dispersion MVDs, therefore assessing the source term was considered important even though the experimental measurements of the source term was not as accurate as in release-specific tests. Additionally, no jet releases from fluids stored in gaseous form were included since, in most cases, source term models implement the formulas for gas flow through an orifice (subsonic or choked) and can be validated against those equations.

Model Acceptance Criteria

In the flammable and toxic MEPs previously developed as part of this project, the quantitative evaluation of a model was based on the definition of a set of statistical performance parameters (e.g., mean relative bias and square error, geometric mean bias and variance, etc.); the values calculated from the simulation of the trials in the MVD were then compared with acceptability criteria, which had previously been established based on the performance of "good models" [5]. A different approach was followed for the quantitative evaluation in the source term MEP, based on the calculation of bias and scatter of the model predictions relative to selected experimental data points. This approach has the advantages of 1) using a quantitative method to determine the validation factor for each model; and 2) allowing the acceptability criteria to be made consistent across different types of hazards and tied to the regulator's own risk tolerance criteria. The quantitative model performance analysis is discussed in detail by Luketa [6] and is summarized below.

Luketa's approach quantifies the model uncertainty and provides performance measures for the degree of over- or under-prediction (quantified by the bias factor, δ) and scatter (quantified by the relative standard deviation, $\tilde{\sigma}_M$), under the assumption that model predictions are normally distributed about true values multiplied by the bias factor.

The model uncertainty can be quantified if both the measurement uncertainty and the propagated input uncertainty (the uncertainty in model parameter inputs due to uncertainty in experimental measurements of those parameters) – together, the "combined uncertainty" ($\tilde{\sigma}_E$) – are known. The formula for the model uncertainty is:

$$\tilde{\sigma}_M^2 + \tilde{\sigma}_E^2 \approx \frac{1}{n-1}\sum_{i=1}^{n}\left[ln\,(M_i/E_i) - \overline{ln\,(M/E)}\right]^2 \qquad (2)$$

Where:

- E_i are individual measurements; and
- M_i are the respective model predictions

The model bias factor (δ) can then be calculated by:

$$\delta = exp\left(\overline{ln\,(M/E)} + \frac{\tilde{\sigma}_M^2}{2} - \frac{\tilde{\sigma}_E^2}{2}\right) \qquad (3)$$

Assuming, given the scope of this MVD, that mass flow rate is the variable of interest, probability that the "true" model's prediction ($\dot{m}$) exceeds a target mass flow rate ($\dot{m}_c$) can be expressed as:

$$\mathrm{P}(\dot{m} > \dot{m}_c) = \frac{1}{2}\mathrm{erfc}\left(\frac{\dot{m}_c - \mu}{\sigma\sqrt{2}}\right) \qquad (4)$$

Where:

- $\mu = {}^{M}/_{\delta}$ is the mean for a normal distribution
- $\sigma = \mu \cdot \tilde{\sigma}_M$ is the standard deviation for the same distribution

Once the probability of exceedance (the probability that the model's prediction will exceed the target value) is specified, equation (4) can be used to calculate the "validation factor" – that is, the factor by which the model predictions need to be multiplied in order for the results to satisfy the probability of exceedance requirement.

As an example, consider the data shown in Figure 1: the red dots indicate the experimental measurements for the mass flow rate through an orifice, as the system pressure is varied, while the blue dots indicate the model's predictions for the same pressures; the error bars indicate the experimental and propagated input uncertainties.

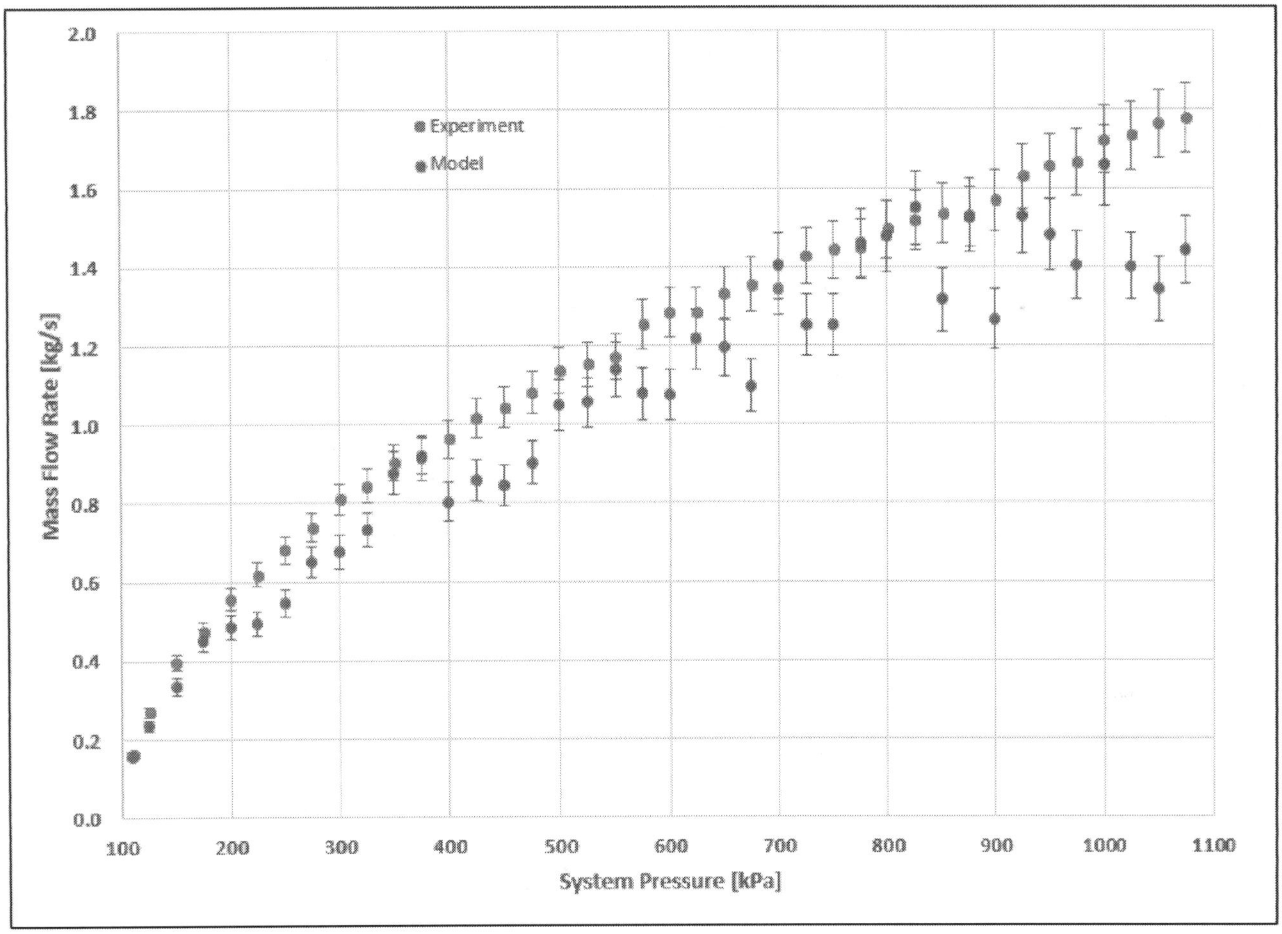

Figure 1. Experimental measurements (red) and model predictions (blue) for the sample dataset.

It is evident from a qualitative inspection of Figure 1 that the model is generally underpredicting the experimental measurements. Assuming a 5% combined uncertainty, the model bias and relative standard deviation can be calculated, respectively, as 0.89 and 0.065; the quantitative assessment confirms that the model is under-predictive (bias < 1). When calculating the mass flow rate from a release, a conservative (over-predictive) model is preferrable, therefore, a high probability of exceedance is desirable: a value of 99% was chosen for consistency with the fire MEP [6]. If the target mass flow rate is set to 1 kg/s, the model validation factor for the example above is equal to 1.29, meaning that the model predictions need to be multiplied by 1.29 to obtain the "true" values to the specified probability of exceedance.

The probability of exceedance is a critical parameter when evaluating a model and would therefore require serious consideration. Figure 2 compares the experimental data with the "true" model predictions (i.e., the model predictions multiplied by the validation factor): the model is now overpredicting the experimental data at all but one point, consistent with the 99% probability of exceedance. The "true" model, however, would not be considered to fit the data well in a traditional model evaluation.

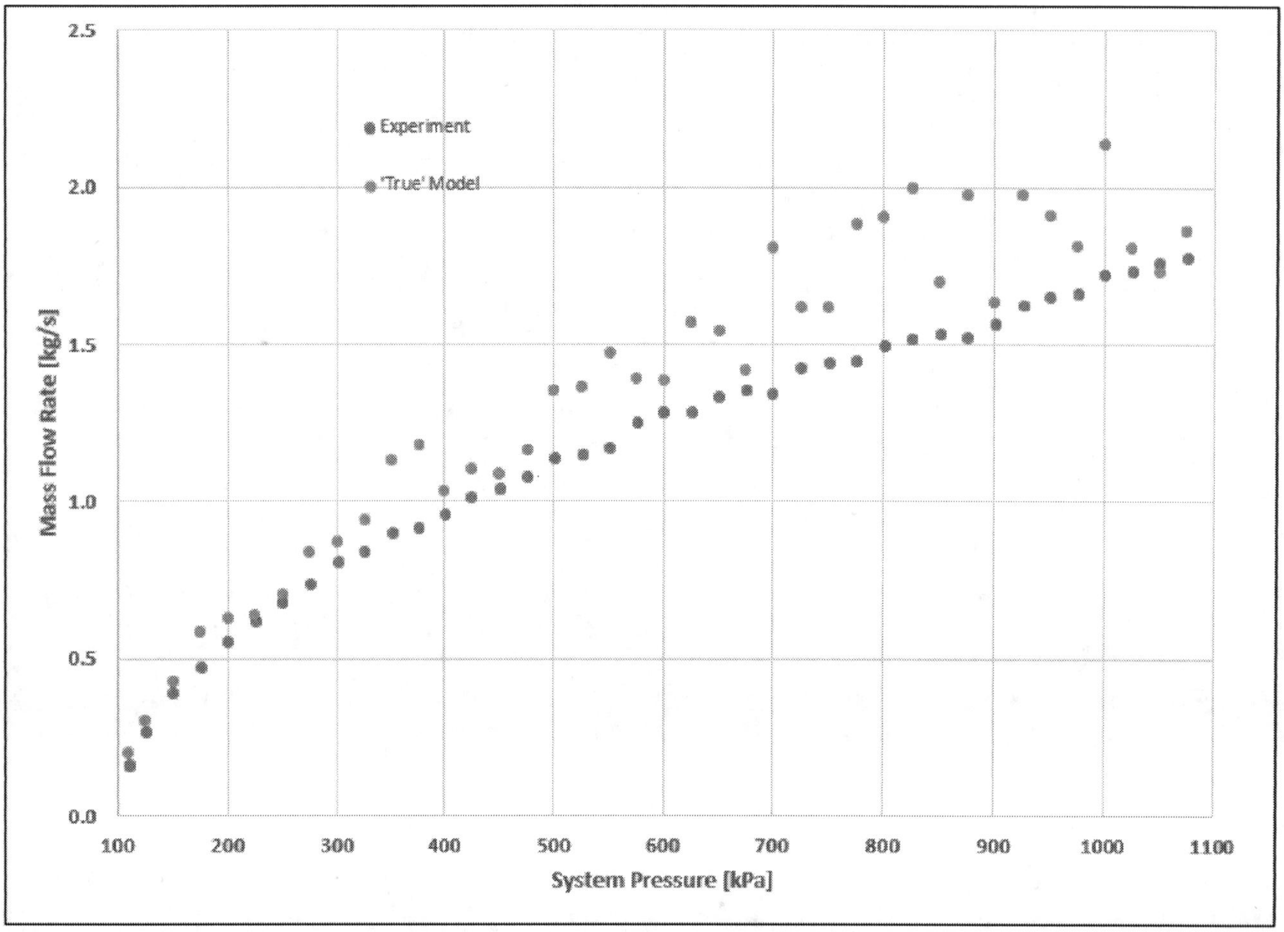

Figure 2. Experimental measurements (red) and "true" model predictions (green) for the sample dataset.

A best fit analysis can be performed for the data shown in Figure 1, and the resulting model is shown in Figure 3 (yellow dots with black edges). The "best fit" model was obtained by applying a validation factor of 1.10.

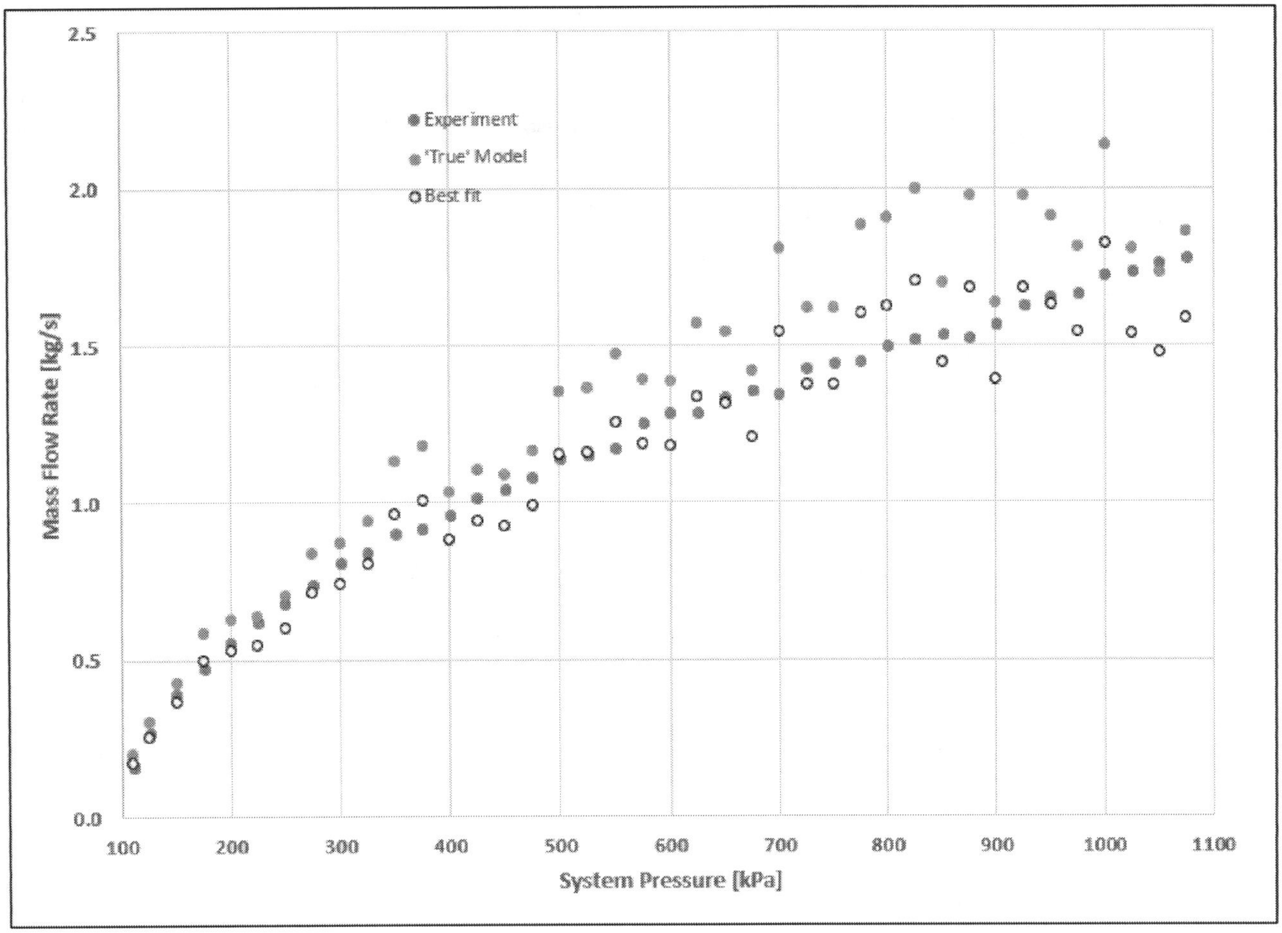

Figure 3. Experimental measurements (red), "true" model predictions (green) and "best fit" model predictions (yellow) for the sample dataset.

For the purpose of the source term MEP, the probability of exceedance is currently set at 99%, to maintain consistency with the fire MEP; however, a technical discussion among stakeholders (e.g., regulators, subject matter experts, etc.) should occur to determine whether this threshold is appropriate or should be changed.

Conclusions

This paper described the scope of the Model Evaluation Protocol for source term models; this is the third in a set of five protocols aimed at providing PHMSA (as well as other regulatory bodies that may choose to follow the same approach) with the tools to evaluate models for the different hazards associated with the siting of LNG facilities.

The paper provided a summary of the main section in each MEP as well as a list of the test series included in the model validation database. The readers were directed to a public website which hosts the full MEP and MVD reports for further information.

The paper also discussed a different approach to model evaluation from the previous MEPs; the current approach provides a more quantitative and objective evaluation of a model's performance and calculation of a validation factor; however, the validation factor is highly dependent on the specified probability of exceedance, which should be selected based on a technical discussion among stakeholders.

References

[1] T. Skjold *et al.*, "A Matter of Life and Death: Validating, Qualifying and Documenting Models for Simulating Flow-Related Accident Scenarios in the Process Industry," *Chem. Eng. Trans.*, vol. 31, 2013.

[2] F. Gavelli, B. Hendrickson, C. Marsegan, J. Piekarz, R. Kooy, and B. Kolade, "Model Validation Database for Flammable Releases," Pipeline and Hazardous Materials Safety Administration, 03903-RP-003, Aug. 2020.

[3] F. Gavelli, B. Hendrickson, C. Marsegan, J. Piekarz, R. Kooy, and B. Kolade, "Model Validation Database for Toxic Releases," Pipeline and Hazardous Materials Safety Administration, 03903-RP-004, Dec. 2020.

[4] F. Gavelli, B. Hendrickson, J. Kolade, and R. Kooy, "Towards a comprehensive model evaluation protocol for LNG hazard analyses," presented at the 23rd Annual Process Safety International Symposium, College Station, TX, Oct. 2020, p. 14.

[5] J. C. Chang and S. R. Hanna, "Air quality model performance evaluation," *Meteorol. Atmospheric Phys.*, vol. 87, no. 1–3, Sep. 2004, doi: 10.1007/s00703-003-0070-7.

[6] A. Luketa, "Model Evaluation Protocol for Fire Models Involving Fuels at Liquefied Natural Gas Facilities.," SAND2022-6812, May 2022. doi: 10.2172/1869376.

37c Identification and Assessment of Risk Factors in Offshore Wind-Integrated Hydrogen Production Systems

Ahmed Elkady, Md Tanjin Amin, Faisal Khan, Mahmoud El-Halwagi

Green hydrogen production through water splitting is a clean energy intensive process that intakes fresh water and electricity from renewables and emits oxygen as a sole byproduct. Recently, due to soaring fossil fuel prices and highly concerning climate change effects, green hydrogen has acquired increasing attention as a potential to decarbonize hard-to-electrify industries and support the growth of renewables. At offshore conditions, water is freely available from seawater, however, appropriate desalination and deionization is necessary to reach the water quality of conventional electrolysis. Additionally, electrolysis plant can be connected to offshore wind farm to transfer (fraction of) its power into hydrogen for storage to be used as fuel or converted back to electricity by means of fuel cell.

In current work, we are seeking to identify the risk factors involved in offshore wind-integrated centralized green hydrogen production system. Risk factors can be related to the technology, economics, safety and security, environment and logistics. Technology risks involve (i) technological and commercial readiness related risks, (ii) availability of resources, (iii) operation risks related to electrolyzer stack operating conditions (e.g., intermittent power, gas crossover, membrane degradation, etc.), water treatment operating conditions (e.g., SWRO membrane fouling), and most importantly risk from hydrogen transportation and storage (e.g., hydrogen leakage, contamination, embrittlement, etc.), and (iv) control related risks. Economics risks involve the factors that determine the economical validity of the whole process. Such risks can be related to (i) capital and operating costs, (ii) macro-economic such as inflation and interest rate risks, and (iii) regulations related to taxation policies such as Investment Tax Credit (ITC) and Production Tax Credit (PTC). Safety and security of the operations and processes is determined by the frequency of accidents and cyber-security breaches and the resilience of the system towards such hazards. Environmental risks involve (i) ecological sensitivity effects from procedures such as desalination wastewater disposal, and (ii) natural hazards such as hurricanes, pandemics, heavy rain, and high seas. Logistical risks mainly involve risks that arise during transporting personnel, supplies as well as during emergency response. In the current work, a thorough investigation has been performed to identify the risk factors as well as their contextual interrelationships, followed by a quantitative analysis of these factors, which yields an estimate of the overall project risk. The current work will be a key decision-making tool for decision-makers to establish offshore water electrolysis-based hydrogen production plants.

37d Development of dynamic process simulator on Fuel Gas Supply System for LNG fueled ship

Akira Tanoue, Kenichi Matsuoka, David Hill

LNG becomes increasingly viable and popular as a marine fuel. Because LNG is clean burning gas, fewer NOX emissions, and has the lowest emissions level per joule of energy released of all fossil fuels.

The use of LNG on board is attracting to comply with strict emission rules by the International Maritime Organization (IMO).

Fuel gas supply system (FGSS) for LNG fueled ships is a fuel gas handling system, and LNG fuel is delivered to the engine and boiler system for propulsion and power generation of vessels.

This study conducted dynamic simulation using the system of CHEMCAD simulator and DCS, to analyze the control stability of fuel gas supply system (FGSS), when FGSS start-up/shutdown, LNG fuel consumption is changing with engine demand, and gas trip /crash astern etc.

The dynamic process simulations for the above scenario cases were conducted to verify the behavior of controllability of fuel gas supply pressure from FGSS to engine system, and the interactions of BOG compressor control and LNG pump control etc.

Additionally, based on the above study, control parameters were tuned before GAS trial operation (the actual plant commissioning) in advance.

AUTHOR INDEX